THE SUN'S GENESIS

Terrence Howard's Theory
of Planetary Origins

Oteren.Fredrick

DISCLAIMER

The substance introduced in this book mirrors the investigation of Terrence Howard's hypothesis that the Sun is the beginning, all things considered, and is expected for enlightening and instructive purposes as it were. The theories and ideas talked about thus depend on the ongoing comprehension of the hypothesis and its suggestions, and may incorporate understandings that are speculative or not all around acknowledged inside mainstream researchers.

COPYRIGHT

© [2024] [Oteren.Fredrick]. Protected by copyright law.

No piece of this distribution might be replicated, dispersed, or sent in any structure or using any and all means, including copying, recording, or other electronic or mechanical techniques, without the earlier composed consent of the creator, with the exception of brief citations epitomized in basic surveys and certain other noncommercial purposes allowed by intellectual property regulation.

TABLE OF CONTENT

INTRODUCTION
CHAPTER 1:
- The Science of Stellar Formation
- 1.1 The Introduction of a Star
- 1.2 The Protostellar Stage
- 1.3 Heavenly Advancement and Principal Sequence
- 1.4 Planetary Arrangement in the Protoplanetary Disk
- 1.5 Heavenly Lifecycles and the Advancement of Planetary Systems
- 1.6 Current Getting it and Observations
- 1.7 The Standard Model versus Elective Theories

CHAPTER 2:
- Terrence Howard's Breakthrough
- 2.1 Terrence Howard: From Entertainer to Theorist
- 2.2 The Beginning of Howard's Theory
- 2.3 Reasonable Structure of Howard's Theory
- 2.4 Howard's Motivation and Influences
- 2.5 Difficulties and Criticisms
- 2.6 Help and Exploration
- 2.7 Ramifications of Howard's Theory

CHAPTER 4:
- Evidence and Counterarguments

- [4.1 Supporting Proof for Howard's Theory](#)
- [4.2 Counterarguments Against Howard's Theory](#)
- [4.3 Hypothetical and Observational Challenges](#)

CHAPTER 5:

- [Implications and Future Research](#)
- [5.1 Ramifications of Howard's Theory](#)
- [5.2 Future Exploration Directions](#)
- [5.3 Assessing Howard's Hypothesis in Context](#)

CHAPTER 6:

- [Broader Impact](#)
- [6.1 Effect on Logical Fields](#)

- **6.2 Innovative and Systemic Advancements**
- **6.3 Effect on Logical Schooling and Public Perception**

CHAPTER 7:
- **Comparative Theories**
- **7.1 The Protoplanetary Plate Model**
- **7.2 The Nebular Hypothesis**
- **7.3 The Center Gradual addition Model**
- **7.4 The Plate Unsteadiness Model**
- **7.5 The Sun oriented Cloud Theory**

CHAPTER 8:
- **Theoretical Models and Simulations**

- **8.1 Hypothetical Models in Planetary Formation**
- **8.1.1 The Protoplanetary Plate Model**
- **8.1.2 The Center Gradual addition Model**
- **8.1.3 The Circle Insecurity Model**
- **8.2 Reenactments and Computational Approaches**
- **8.3 Assessing Howard's Hypothesis with Simulations**
- **8.4 Future Headings in Demonstrating and Simulation**

CHAPTER 9:
- **Scientific and Philosophical Perspectives**
- **9.1 Logical Perspectives**
- **9.2 Philosophical Perspectives**

- **9.3 The Job of Interdisciplinary Approaches**
- **9.4 Future Philosophical and Logical Inquiry**

CHAPTER 10:
- **The Role of Media and Public Perception**
- **10.1 Media Portrayal of Logical Theories**
- **10.2 Public Impression of Eccentric Theories**
- **10.3 The Effect of Media on Logical Discourse**
- **10.4 Systems for Powerful Science Communication**
- **10.5 Contextual analyses and Examples**
- **10.6 Future Directions**

CONCLUSION

APPENDICES
ACKNOWLEDGEMENTS

INTRODUCTION

The universe has consistently captivated mankind, convincing us to look for replies about the starting points and nature of our universe. Among the bunch speculations that endeavor to make sense of the arrangement of divine bodies, one especially charming suggestion has risen up out of a far-fetched source: entertainer and business person Terrence Howard. Known essentially for his jobs in film and TV, Howard has wandered into the domain of astronomy with a hypothesis that challenges the standard way of thinking about the planetary group's development. His speculation proposes that the actual Sun is the beginning, all things considered, an idea that veers essentially from the broadly acknowledged models in astronomy.

To see the value in the boldness and possible meaning of Howard's hypothesis, it is fundamental for first grasp the customary models of heavenly and planetary arrangement. As per winning logical agreement, stars and planetary frameworks structure through an interaction known as heavenly nucleosynthesis and growth. This cycle starts in immense sub-atomic mists, where districts of higher thickness breakdown under their own gravity, prompting the arrangement of protostars. As the protostar develops, it amasses encompassing material, in the long run touching off atomic combination at its center. This combination makes a star, and the extra material structures a protoplanetary circle. After some time, particles inside this plate impact and mix to frame planets, moons, and other heavenly bodies.

This model, which has been refined through many years of perception and hypothetical turn of events, gives a structure to figuring out the development of our planetary group and others. It makes sense of how the Sun shaped from a haze of gas and residue and how the planets, moons, and different bodies in our planetary group began from the remaining material left over after the Sun's development. Be that as it may, Howard's hypothesis acquaints an extreme contort with this deep rooted story.

Terrence Howard's advantage in science, especially in the fields of physical science and cosmology, has been a longstanding one. His scholastic foundation, however unpredictable contrasted with normal researchers, is set apart by a profound interest and a creative way to deal with critical thinking. Howard's hypothesis suggests that as opposed to framing from a protoplanetary plate, the planets were made straightforwardly from the actual Sun. As per Howard, the Sun isn't simply a latent body at the focal point of the nearby planet group yet a functioning wellspring of planetary material.

Howard's speculation is grounded in the idea of sun based beginning, a cycle he accepts includes the Sun removing material that step by step consolidates into planets. This thought stands out strongly from the standard model, which places that the Sun's development is a different occasion from the development of the planets. In Howard's view, the Sun's energy and material are vital to the making of planetary bodies, proposing a more interconnected connection between the star and the planets it purportedly makes.

The ramifications of Howard's hypothesis are significant. In the event that the Sun for sure assumed an immediate part in the development of planets, it could change how we might interpret heavenly development and the elements of planetary frameworks. This thought difficulties the laid out view that planets structure as a result of the star's growth cycle. All things being equal, it proposes a more unique and coordinated model where the star and planets are essential for a constant pattern of issue and energy trade.

To completely get a handle on the meaning of Howard's hypothesis, one should consider both the potential proof supporting it and the reactions it has confronted. Howard's thoughts are not in view of conventional logical exploration yet rather on a calculated system that tries to reconsider how we figure out sunlight based and planetary development. Thusly, his hypothesis has ignited impressive discussion inside established researchers. A few scientists are charmed by the original viewpoint and are investigating ways of testing Howard's thoughts through reenactments and observational examinations. Others still have doubts, contending that Howard's hypothesis needs observational help and doesn't line up with existing cosmic information.

In analyzing Howard's hypothesis, taking into account the more extensive setting of logical innovation is essential. From the beginning of time, logical advancement has frequently been driven by people who challenge winning standards and proposition new points of view. Howard's hypothesis addresses one such test, pushing the limits of our comprehension and empowering a reconsideration of laid out models. Whether his speculation demonstrates right, it fills in as a sign of the significance of scholarly interest and the eagerness to investigate capricious thoughts.

This book intends to give an exhaustive investigation of Terrence Howard's hypothesis, from its beginnings and components to its suggestions and gathering inside established researchers. We will dive into the subtleties of Howard's speculation, contrasting it and customary models of planetary arrangement and analyzing the proof and scrutinizes related with it. Thusly, we desire to reveal insight into the expected effect of Howard's thoughts and their place inside the more extensive scene of astrophysical exploration.

As we leave on this excursion, it is vital for approach Howard's hypothesis with a receptive outlook, perceiving that logical request is an advancing interaction portrayed by consistent addressing and refinement. The investigation of new speculations, regardless of how whimsical, assumes an essential part in propelling comprehension we might interpret the universe. Howard's hypothesis, with its strong cases and creative viewpoint, addresses an astonishing section in the continuous mission to unwind the secrets of the universe.

In the parts that follow, we will dig further into the logical and reasonable underpinnings of Howard's hypothesis, looking at its suggestions and investigating the continuous discourse between novel thoughts and laid out logical information. Through this investigation, we expect to give a nuanced and adjusted point of view on Howard's commitment to how we might interpret the planetary group and the more extensive universe.

CHAPTER 1:

The Science of Stellar Formation

The development of stars and planetary frameworks is a principal cycle in astronomy, unpredictably attached to the advancement of the universe. To comprehend Terrence Howard's hypothesis that the Sun is the beginning of all planets, we should initially dig into the laid out study of heavenly development. This part will investigate the components by which stars are conceived, develop, and eventually lead to planetary frameworks, making way for a more profound assessment of Howard's speculation.

1.1 The Introduction of a Star

Stars are shaped in immense, cold billows of gas and residue known as sub-atomic mists, or heavenly nurseries. These mists are principally made out of hydrogen, the most plentiful component known to man, alongside helium and follow measures of different components. The course of star development starts when a district inside a sub-atomic cloud encounters an expansion in thickness, frequently set off by outside powers, for example, shock waves from neighboring supernovae or connections with different mists.

As a specific locale of the cloud implodes under its own gravity, it frames a thick center known as a protostar. This center keeps on accumulating material from the encompassing cloud, prompting a climb in temperature and tension. At the point when the center's temperature arrives at roughly 10 million degrees Kelvin, atomic combination responses start, denoting the introduction of another star. The beginning of combination delivers an outward strain that neutralizes the gravitational breakdown, settling the star.

1.2 The Protostellar Stage

During the protostellar stage, the shaping star is encircled by an alternating circle of gas and residue. This protoplanetary plate assumes a urgent part in the ensuing development of planetary frameworks. The plate's material is warmed by the youthful star's radiation and is dependent upon complex actual cycles, including disturbance, attractive fields, and rakish energy move.

The protoplanetary plate is where planetary development starts. Dust particles inside the plate impact and remain together, shaping bigger bodies known as planetesimals. These planetesimals slowly mix to shape protoplanets, which further develop through extra impacts and growth. Over the long run, these protoplanets clear their circles and become undeniable planets.

1.3 Heavenly Advancement and Principal Sequence

When atomic combination is laid out, the star enters the primary arrangement period of its life cycle. During this stage, the star consistently melds hydrogen into helium in its center, delivering the energy that controls the star and gives the outward tension expected to balance gravitational breakdown. This stage can last billions of years, contingent upon the star's mass. The principal grouping stage is described by a harmony between the gravitational powers pulling the star internal and the radiation pressure pushing outward. The star's soundness and life span during this stage are vital for the improvement of any planetary frameworks conformed to it.

1.4 *Planetary Arrangement in the Protoplanetary Disk*

As the star balances out on the primary arrangement, the excess material in the protoplanetary circle keeps on developing. This material goes through different cycles that shape the arrangement of planets and other divine bodies. Key cycles include:

- Accretion: The continuous aggregation of material onto bigger bodies through crashes and gravitational fascination.
- Differentiation: The cycle by which planetesimals and protoplanets separate into various layers in view of thickness, framing center, mantle, and covering structures.

- Getting free from the Disk: The star's radiation and sunlight based breeze in the end blow away the leftover gas in the circle, finishing the planetary arrangement process.

1.5 Heavenly Lifecycles and the Advancement of Planetary Systems

The lifecycle of a star impacts the qualities of planetary frameworks conformed to it. Low-mass stars, similar to our Sun, have long life expectancies and give a steady climate to planetary frameworks to develop. High-mass stars, then again, have more limited life expectancies and frequently end their lives in emotional cosmic explosion blasts, which can fundamentally affect close by planetary frameworks.

The end phases of a star's life cycle, including red monster and cosmic explosion stages, can likewise influence the encompassing planetary framework. For example, the development of a red goliath can prompt the deficiency of planetary environments or even the launch of planets from the framework.

1.6 Current Getting it and Observations

Present day perceptions and recreations have given significant proof supporting the conventional model of heavenly and planetary arrangement. Telescopes like the Hubble Space Telescope and the impending James Webb Space Telescope offer point by point pictures of star-shaping areas and protoplanetary plates, giving important experiences into the cycles in question.

Reproductions of star and planet development assist researchers with understanding the elements of protoplanetary plates and the arrangement of planetary frameworks. These recreations integrate complex physical science, including hydrodynamics, magnetohydrodynamics, and substance processes, to display the development of stars and planets.

1.7 The Standard Model versus Elective Theories

The standard model of heavenly and planetary development is upheld by broad observational and hypothetical examination. Be that as it may, elective speculations, including Terrence Howard's theory, propose various components for planetary development. These elective speculations frequently try to address holes or irregularities in the standard model or to investigate clever thoughts regarding the arrangement and development of heavenly bodies.

Howard's hypothesis, which proposes that the Sun is the beginning of all planets, challenges the customary view that planets structure from a protoplanetary plate. His speculation will be analyzed exhaustively in resulting parts, yet understanding the standard model gives a vital establishment to assessing his thoughts.

The study of heavenly development is a complex and advancing field that consolidates observational cosmology with hypothetical displaying. The conventional model, which portrays stars shaping from sub-atomic mists and making planetary frameworks through protoplanetary circles, has been upheld by broad exploration and proof. As we investigate Terrence Howard's hypothesis that the Sun is the beginning, all things considered, it is fundamental to think about this laid out system and comprehend how groundbreaking thoughts fit inside or challenge existing information.

In the accompanying sections, we will dig into Howard's hypothesis, analyzing its systems, proof, and suggestions with regards to the laid out study of heavenly and planetary development. This investigation will give a complete perspective on how Howard's speculation converges with and wanders from customary models, offering new viewpoints on the beginnings of our nearby planet group and then some.

CHAPTER 2:

Terrence Howard's Breakthrough

Terrence Howard, generally perceived for his work in film and TV, has wandered into the domain of logical hypothesis with a suggestion that has charmed the two researchers and the public the same. Howard's leading edge hypothesis challenges customary comprehension by recommending that the actual Sun is the beginning, all things considered. This section investigates Howard's hypothesis exhaustively, analyzing its turn of events, the logical and reasonable underpinnings, and the setting where it arose.

2.1 Terrence Howard: From Entertainer to Theorist

Terrence Howard's excursion into the universe of logical hypothesis is capricious. Most popular for his parts in films like Hustle and Flow and Iron Man, Howard's introduction to logical talk has accumulated critical consideration. His advantage in science, especially in physical science and cosmology, has been a well established part of his scholarly interests. Howard's scholarly foundation and encounters, while not conventional for a physicist, mirror a profound interest and an eagerness to investigate inventive thoughts.

Howard's association in logical hypothesis started with his interest with major inquiries concerning the universe. He has communicated a strong fascination with understanding the idea of issue and energy, which drove him to foster his own speculation about planetary development. His methodology joins a mix of instinct, reasonable reasoning, and whimsical strategies for request.

2.2 The Beginning of Howard's Theory

Howard's hypothesis proposes an extreme shift from the traditional comprehension of planetary development. As indicated by Howard, the Sun isn't only a focal star around which planets structure, yet rather a functioning wellspring of planetary material. He proposes that the actual Sun is liable for the making of planets, generally modifying our impression of the planetary group's beginnings.

Howard's speculation can be summed up as follows: rather than shaping from a protoplanetary circle, planets are created straightforwardly from the Sun. This cycle includes the Sun radiating material that consolidates into planetary bodies. Howard portrays this as a type of sun powered beginning, where the Sun assumes an immediate and dynamic part in the development of planets.

2.3 Reasonable Structure of Howard's Theory

Howard's hypothesis is based on a few key ideas:

- Sun powered Genesis: Howard suggests that the Sun, through a course of material ejection or change, makes planets. This idea challenges the conventional view that planets structure from a remaining plate of material encompassing a recently framed star.
- Material Emission: As per Howard, the Sun transmits matter in a way that permits it to combine into planetary bodies. This discharge could include complex communications between sun oriented radiation, attractive fields, and material launch.
- Dynamic Interaction: Howard imagines a unique collaboration between the Sun and the shaping planets. Instead of an inactive interaction where planets gradually structure from a circle, Howard's model recommends a functioning trade of material and energy.

2.4 Howard's Motivation and Influences

Howard's hypothesis is affected by different logical and philosophical thoughts. His experience in artistic expression and amusement has molded his imaginative way to deal with critical thinking, which he applied to his logical speculation. Howard has communicated reverence for crafted by spearheading researchers and rationalists, and his hypothesis mirrors a blend of these impacts.

Howard's thoughts are likewise propelled by the perception of heavenly peculiarities and hypothetical ideas from astronomy. While his hypothesis wanders from standard science, it mirrors a more extensive interest in understanding the essential cycles that oversee the universe.

2.5 Difficulties and Criticisms

Howard's hypothesis has confronted critical examination from established researchers. Key difficulties include:

- Absence of Observational Evidence: One significant analysis is the absence of exact proof supporting Howard's speculation. The customary model of planetary development is upheld by broad observational information and recreations, while Howard's hypothesis depends more on reasonable thoughts.
- Hypothetical Consistency: Pundits contend that Howard's hypothesis might need consistency with laid out standards of physical science and astronomy. The proposed instruments for material outflow and planetary development should be thoroughly tried against known actual regulations.
- Examination with Existing Models: Howard's hypothesis should be contrasted and the laid out model of heavenly development, which has been

approved through various perceptions and studies. This correlation features possible errors and regions for additional examination.

2.6 Help and Exploration

Notwithstanding the reactions, Howard's hypothesis has ignited interest and investigation inside mainstream researchers. A few specialists are fascinated by the original viewpoint and are investigating ways of testing the hypothesis through reenactments and observational examinations. These endeavors expect to assess whether Howard's thoughts offer new experiences or require critical amendment.

2.7 Ramifications of Howard's Theory

In the event that Howard's hypothesis were approved, it could have significant ramifications for how we might interpret the planetary group and planetary science. Key ramifications include:

- Reconsideration of Heavenly and Planetary Formation: The hypothesis would require a reexamination of how stars and planets structure, possibly prompting new models and speculations in astronomy.

- New Viewpoints on Sun powered Dynamics: Howard's speculation could give new bits of knowledge into the elements of the Sun and its association with encompassing material.

- Effect on Future Research: The hypothesis might motivate new examination headings and techniques, provoking researchers to investigate flighty thoughts and test novel speculations.

Terrence Howard's cutting edge hypothesis presents a strong and capricious point of view on the beginnings of planets. By recommending that the actual Sun is the wellspring of planetary material, Howard challenges laid out models of heavenly and planetary development. His hypothesis mirrors an inventive and creative way to deal with figuring out the universe, pushing the limits of customary science.

As we keep on investigating Howard's speculation, moving toward it with both interest and basic scrutiny is fundamental. The investigation of novel thoughts, even those that challenge laid out information, is a fundamental piece of logical advancement. Howard's hypothesis welcomes us to reexamine the idea of planetary development and to investigate the potential outcomes that emerge from reconsidering how we might interpret the universe.

In the accompanying parts, we will dive further into the components of Howard's hypothesis, analyze the proof supporting it, and think about the more extensive ramifications for astronomy and planetary science. Through this investigation, we expect to acquire an exhaustive comprehension of Howard's commitment to the field and its likely effect on our insight into the universe.

CHAPTER 3:

The Mechanics of Howard's Theory

Terrence Howard's hypothesis places a progressive thought: that the actual Sun is the beginning, all things considered. This speculation remains as an unmistakable difference to the generally acknowledged model of planetary development, which depicts planets shaping from a protoplanetary plate around a youthful star. To completely comprehend Howard's recommendation, it is significant to dig into the mechanics of his hypothesis — how he imagines the Sun delivering planets and the cycles in question.

3.1 The Idea of Sunlight based Genesis

Howard's hypothesis presents the idea of "sun oriented beginning," which proposes that the Sun isn't only a detached base on which planets structure however a functioning wellspring of planetary material. As indicated by this thought, the Sun produces planets straightforwardly from its own substance through a course of material removal or change. This idea essentially challenges the customary view that planets structure from a remaining plate of gas and residue encompassing a recently framed star.

3.2 Instruments of Planetary Emission

Howard's speculation includes a few proposed systems by which the Sun could emanate material that in the long run shapes planets:

- Material Ejection: Howard proposes that the Sun could oust material as sun powered flares or coronal mass launches. These occasions discharge huge amounts of issue into space, which could, as per the hypothesis, blend into planetary bodies. This component suggests that the Sun's movement straightforwardly adds to the arrangement of planets.

- Sun oriented Breeze and Attractive Fields: One more proposed component includes the collaboration of the's sun powered breeze and attractive fields with encompassing space. Howard speculates that the sun oriented breeze could divert material from the Sun, which, impacted by attractive fields and different powers, could slowly frame planets. This cycle would include complex elements between the Sun's result and the general climate.

- Inner Processes: Howard additionally investigates the possibility that inward cycles inside the Sun could add to planetary development. This could incorporate changes inside the Sun's center or external layers that outcome in the ejection of material equipped for shaping planets. The particulars of these cycles stay speculative and would require further examination.

To evaluate Howard's hypothesis, contrasting it and the conventional model of planetary formation is fundamental:

- Protoplanetary Plate Model: The broadly acknowledged model of planetary development includes a protoplanetary circle, an alternating plate of gas and residue encompassing a recently framed star. Planets structure through growth, where dust particles impact and remain together to frame bigger bodies. Over the long haul, these bodies develop into planets, moons, and other heavenly items.

- Howard's Model: conversely, Howard's model proposes that the actual Sun is the wellspring of planetary material. This model doesn't depend on a protoplanetary circle however rather recommends that the Sun's own material, through different cycles, is ousted or changed into planets. This view requires a reconsideration of how we figure out the development and advancement of planetary frameworks.

Howard's hypothesis faces a few hypothetical and experimental difficulties:

- Consistency with Actual Laws: One significant test is guaranteeing that Howard's hypothesis lines up with laid out actual regulations. The customary model of planetary development is all around upheld by observational information and hypothetical exploration. Howard's hypothesis should be predictable with known standards of material science, including protection of mass and energy, and the elements of heavenly and planetary frameworks.

- Experimental Evidence: Another test is the absence of exact proof supporting Howard's hypothesis. The conventional model is upheld by broad perceptions of protoplanetary circles, gradual addition processes, and the development of planetary frameworks. Howard's hypothesis requires new proof or reevaluation of existing information to acquire more extensive acknowledgment.

- Observational Data: Howard's hypothesis should likewise be assessed considering current observational information. Telescopes like the Hubble Space Telescope and the James Webb Space Telescope have given definite pictures and information on star-shaping locales and protoplanetary plates. Howard's model should be tried against these perceptions to decide its legitimacy.

Assuming Howard's hypothesis were approved, it could have a few huge ramifications:

- Reexamined Models of Planetary Formation: An approved hypothesis of sunlight based beginning could prompt modified models of planetary development, integrating new instruments and cycles. This would require a reexamination of how planetary frameworks are shaped and the way that they develop after some time.

- Understanding Sun oriented Dynamics: Howard's hypothesis could give new bits of knowledge into the elements of the Sun and its communication with encompassing material.

This could improve how we might interpret sun based action, including sun powered flares, coronal mass discharges, and their effect on planetary development.

- Impact on Astrophysical Research: The hypothesis could rouse new lines of examination and trial and error in astronomy. It might incite researchers to investigate capricious thoughts and test new speculations, adding to the progression of the field.

To additionally explore Howard's hypothesis, a few exploration headings could be sought after:

- Recreation and Modeling: Creating reproductions and models to test the instruments proposed by Howard's hypothesis. This could include displaying sun based material discharge, attractive field collaborations, and the development of planetary bodies from ousted material.

- Observational Studies: Leading observational examinations to look for proof of sun oriented material discharge and its likely job in planetary development. This could incorporate checking sun powered action and dissecting the structure of material ousted by the Sun.

- Hypothetical Development: Further fostering the hypothetical system of Howard's hypothesis to guarantee consistency with laid out actual regulations and standards. This could include coordinated effort with physicists and astrophysicists to refine the hypothetical parts of the speculation.

CHAPTER 4:

Evidence and Counterarguments

The assessment of Terrence Howard's hypothesis — that the actual Sun is the beginning of all planets — requires an exhaustive assessment of the proof supporting and testing the speculation. This section dives into the accessible proof that could uphold Howard's thoughts, addresses the counterarguments raised by mainstream researchers, and talks about the hypothetical and observational difficulties looked by his hypothesis.

4.1 Supporting Proof for Howard's Theory

To evaluate the legitimacy of Howard's hypothesis, it is vital to consider any proof that could uphold the thought of sun based beginning:

- Sun oriented Action Observations: Sun powered movement, for example, sun based flares and coronal mass discharges, includes the launch of material from the Sun. While this material normally comprises of charged particles as opposed to strong matter, a few defenders recommend that these launches might actually add to planetary development under unambiguous circumstances. Be that as it may, how much material catapulted is for the most part excessively scanty and spread to frame planets.

- Sythesis of the Sun oriented System: The comparability in arrangement between the Sun and the planets could appear to help the possibility that planets began from the Sun. In any case, this likeness is predictable with the protoplanetary circle model, which places that the Sun and planets share normal material because of their development from a similar sub-atomic cloud.

- Astrophysical Models: Certain astrophysical models investigate the elements of material removed from stars. For example, a few models inspect how material from supernovae adds to the development of new stars and planets. While these models don't straightforwardly uphold Howard's hypothesis, they in all actuality do explore related ideas of material exchange in heavenly frameworks.

4.2 Counterarguments Against Howard's Theory

Howard's hypothesis has confronted a few critical counterarguments from established researchers:

- Experimental Evidence: The customary model of planetary development is upheld by broad observational proof, including the disclosure of protoplanetary circles around youthful stars. Perceptions from telescopes like Hubble and the impending James Webb Space Telescope have given obvious proof of circles from which planets structure. Howard's hypothesis needs direct observational help that can validate the possibility of planets framing straightforwardly from the Sun.

- Actual Constraints: Howard's recommendation that sun based material is straightforwardly launched out to shape planets faces actual limitations. The Sun's launches, for example, sun oriented flares and coronal mass discharges, basically discharge plasma and particles, not strong material. How much material associated with these cycles is moderately little contrasted with what might be expected to shape planets. Furthermore, the actual components by which sun based material could combine into planets are not clear cut or upheld by existing models.

- The Protoplanetary Plate Model: The protoplanetary circle model makes sense of the development of planets with impressive achievement. Perceptions of youthful star frameworks with protoplanetary plates areas of strength for give to this model. Planets shaping from a circle of gas and residue around a star is a deeply grounded process upheld by reproductions and observational information. Howard's hypothesis should address the legitimate cycles of plate arrangement and planetary growth.

- Heavenly Evolution: The customary model likewise consolidates the idea of heavenly development, where the Sun's job in planetary arrangement is uninvolved. The material that structures planets is leftover material from the star-framing process, not created straight by the star. This model lines up with perceptions of other star frameworks and the existence patterns of stars.

4.3 Hypothetical and Observational Challenges

A few difficulties convolute the acknowledgment of Howard's hypothesis:

- Consistency with Actual Laws: Howard's hypothesis should be accommodated with laid out actual regulations, like protection of mass and energy. The conventional model of planetary arrangement lines up with these standards, while Howard's hypothesis includes instruments that are not surely known or tried.

- Absence of Robotic Detail: Howard's hypothesis proposes components for the Sun discharging material that structures planets, yet these instruments are not completely nitty gritty or upheld by hypothetical models. The shortfall of an unmistakable, testable interaction makes it challenging to assess the plausibility of the hypothesis.

Observational Data: There is presently no direct observational proof of planets shaping from sun based material. Perceptions of planetary development center around protoplanetary plates and the growth processes happening inside them. Howard's hypothesis would require new sorts of perceptions or reevaluations of existing information to acquire observational help.

Howard's hypothesis has ignited banter among researchers and specialists:

- Premium in Clever Ideas: A few scientists are keen on investigating unusual hypotheses and may examine parts of Howard's thoughts further. This investigation could include hypothetical displaying or observational examinations to test new theories about heavenly and planetary arrangement.

- Basic Review: Numerous researchers have fundamentally investigated Howard's hypothesis, featuring its deviations from laid out models and the absence of exact proof. Basic audits help refine and challenge logical thoughts, adding to the thorough assessment process that is fundamental in logical talk.
- Potential for Collaboration: The hypothesis has provoked conversations about the potential for interdisciplinary coordinated effort. Analysts from various fields, including astronomy, material science, and observational stargazing, may team up to investigate the achievability of elective hypotheses and components proposed by Howard.

CHAPTER 5:

Implications and Future Research

Terrence Howard's hypothesis — that the actual Sun is the beginning of all planets — offers an extreme takeoff from laid out models of planetary development. Understanding the ramifications of this hypothesis and illustrating expected roads for future examination are vital for assessing its effect on the field of astronomy. This part investigates the more extensive ramifications of Howard's speculation and recommends headings for additional examination.

5.1 Ramifications of Howard's Theory

On the off chance that approved, Howard's hypothesis could have huge ramifications for how we might interpret the planetary group and planetary arrangement:

- Reconsidered Models of Planetary Formation: Howard's hypothesis could prompt the improvement of new models of planetary development that integrate sun powered beginning. This would include reevaluating the systems by which planets structure and develop, possibly coordinating ideas from Howard's speculation with conventional models. Such modified models would have to address both hypothetical and observational parts of planetary development.

- Understanding Sun based Dynamics: The hypothesis could give new bits of knowledge into the elements of the Sun, remembering its job for molding planetary frameworks. On the off chance that the Sun is effectively associated with framing planets, it could adjust how we might interpret sun based action, including the impacts of sun oriented flares, coronal mass discharges, and the sun powered breeze on planetary development.

- Influence on Astrophysical Research: Howard's hypothesis could motivate new lines of exploration in astronomy, provoking researchers to investigate eccentric thoughts and test novel speculations. This could prompt a more extensive investigation of heavenly and planetary development processes, possibly uncovering new parts of the universe.

- Effect on Training and Public Perception: The presentation of another hypothesis like Howard's could impact instructive educational plans and public comprehension of stargazing. It would energize conversations about the idea of logical speculations and the significance of proof in approving speculations.

5.2 Future Exploration Directions

To completely assess Howard's hypothesis and its suggestions, a few areas of examination warrant investigation:

- Hypothetical Turn of events and Modeling: Creating hypothetical models to mimic the proposed components of sun oriented beginning is fundamental. Scientists ought to make models to test the attainability of Howard's thoughts, for example, material launch and its expected job in shaping planets. These models ought to integrate laid out standards of physical science and contrast results and observational information.

- Observational Studies: Directing observational examinations to look for proof of sun oriented material discharge and its expected job in planetary development is vital. This incorporates checking sun based movement and investigating the creation of material catapulted by the Sun. High level telescopes and space missions could give significant information to testing Howard's theory.

- Near Examination with Conventional Models: Scientists ought to play out a relative investigation between Howard's hypothesis and the customary protoplanetary circle model. This examination would include assessing how well each model makes sense of noticed peculiarities and which model offers a more exact portrayal of planetary development.

- Exploratory Testing: Trial draws near, like lab reenactments of sun based material elements, could assist with testing parts of Howard's hypothesis. While direct trial and error on a sun oriented scale is testing, limited scope examinations could give bits of knowledge into the cycles engaged with material launch and growth.
Interdisciplinary Collaboration: Coordinated effort between physicists, space experts, and different researchers is urgent for assessing Howard's hypothesis. Interdisciplinary exploration could unite different viewpoints and aptitude to address the hypothetical and observational difficulties presented by the speculation.

- Long haul Checking and Information Collection: Long haul observing of sun based movement and planetary development processes is fundamental for social occasion complete information. This continuous exploration could assist with distinguishing any examples or peculiarities that could support or challenge Howard's hypothesis.

5.3 Assessing Howard's Hypothesis in Context

Howard's hypothesis should be assessed inside the more extensive setting of astrophysical examination. This includes:

- Incorporating New Findings: As new information and perceptions arise, they ought to be coordinated into the assessment of Howard's hypothesis. This remembers considering late revelations for heavenly and planetary science and how they line up with or go against the proposed speculation.

- Refreshing Logical Models: Logical models and speculations ought to be refreshed to reflect new discoveries. Assuming Howard's hypothesis acquires support, it could prompt updates in how we might interpret planetary development and the job of the Sun.

- Drawing in with the Logical Community: Connecting with established researchers through gatherings, distributions, and conversations is imperative for propelling exploration and cultivating joint effort. Sharing discoveries and getting criticism from companions can help refine and test the hypothesis.

Creative methodologies and advances could assume a critical part in propelling exploration on Howard's hypothesis. This incorporates:

- High level Simulations: Using progressed computational procedures and reenactments to show complex collaborations between the Sun and encompassing material. These reenactments could investigate situations not yet thought of and give new bits of knowledge into sun powered beginning.
- Cutting edge Observatories: Utilizing cutting edge observatories and space missions to acquire high-goal information on sun oriented action and

planetary arrangement. These advances could offer new viewpoints on the cycles proposed by Howard's hypothesis.

- Public Commitment and Education: Connecting with general society and instructive organizations in conversations about new hypotheses and revelations. This can assist with advancing logical proficiency and energize interest in creative exploration.

CHAPTER 6:

Broader Impact

Terrence Howard's hypothesis, recommending that the actual Sun is the beginning of all planets, can possibly impact different fields past conventional astronomy. This section investigates the more extensive effect of Howard's hypothesis, inspecting its suggestions for related logical disciplines, innovation, and society at large.

6.1 Effect on Logical Fields

- Astronomy and Planetary Science: Howard's hypothesis challenges laid out models of planetary development and could prompt a reassessment of how we might interpret star and planet development. Whenever approved, it would provoke huge changes in astrophysical models, requiring a reassessment of how planetary frameworks develop and how they are noticed. This hypothesis could likewise move new examination in planetary science, possibly prompting disclosures about the idea of planetary arrangement and the elements of planetary groups.

- Sunlight based Physics: The hypothesis proposes an immediate connection between sun powered action and planetary development. This could prompt new examinations concerning sunlight based physical science, especially the way in which sun powered peculiarities could impact or add to the development of planetary bodies. Scientists would have to investigate the systems by which the Sun could radiate material fit for framing planets, possibly uncovering new parts of sun oriented elements and action.

- Hypothetical Physics: Howard's theory includes eccentric instruments for material launch and planetary arrangement. This could animate examination in hypothetical physical science, especially in regions connected with material elements, energy move, and the key cycles administering heavenly frameworks. It could prompt new hypothetical structures or adjustments of existing ones.

6.2 *Innovative and Systemic Advancements*

- Observational Technology: To research Howard's hypothesis, progressions in observational innovation would be fundamental. This incorporates the advancement of additional delicate telescopes and instruments equipped for recognizing and dissecting material launched out from stars. Advancements in space-based observatories and ground-based telescopes could give the information expected to test new hypotheses and refine how we might interpret heavenly and planetary cycles.

- Computational Models: The hypothesis might drive headways in computational displaying, requiring recreations that consolidate new components for material discharge and planetary arrangement. Superior execution registering assets and complex displaying strategies could be created to investigate Howard's thoughts and evaluate their achievability.

- Trial Techniques: While direct trial and error on a sun based scale is testing, exploratory strategies could be adjusted to test related ideas on a more limited size. Research facility recreations and trials could offer experiences into the cycles engaged with material discharge and its likely job in planetary development.

6.3 Effect on Logical Schooling and Public Perception

- Instructive Curricula: Howard's hypothesis could impact instructive educational plans in stargazing and astronomy. It might prompt the consideration of new speculations and elective viewpoints in instructive projects, empowering understudies to investigate and basically assess unusual thoughts. This can improve's comprehension understudies might interpret the logical interaction and encourage decisive reasoning.

- Public Engagement: The presentation of another hypothesis like Howard's can draw in the general population in conversations about logical revelation and the idea of logical speculations. It offers a valuable chance to advance logical education and interest in space investigation. Public talks, media inclusion, and effort projects can assist with making sense of the meaning of new hypotheses and the significance of proof based science.

- Science Communication: The more extensive effect of Howard's hypothesis features the requirement for compelling science correspondence. Researchers and teachers should pass complex thoughts and examination discoveries on to general society in a reasonable and open way. This can assist with demystifying logical cycles and encourage a superior comprehension of how new speculations are assessed and approved.

- Cross-Disciplinary Research: Howard's hypothesis energizes interdisciplinary coordinated effort between astrophysicists, sun powered physicists, hypothetical physicists, and other logical fields. Such coordinated efforts can unite different skill and viewpoints, prompting a more complete examination of the proposed components and their suggestions.

- Cooperative Projects: Joint exploration projects including various disciplines could be started to investigate Howard's hypothesis. Cooperative endeavors between research organizations, colleges, and space offices could work with the advancement of new models, observational systems, and exploratory procedures.

- Combination with Arising Technologies: The hypothesis might drive the mix of arising advancements in research. For instance, headways in man-made consciousness and AI could be applied to examine enormous datasets, distinguish examples, and test hypothetical expectations. This coordination could improve our capacity to investigate and approve new logical thoughts.

- Motivation for Innovation: Imaginative speculations like Howard's can move inventiveness and development across different fields. The investigation of whimsical thoughts can prompt surprising disclosures and mechanical progressions that benefit society. The fervor and interest created by new logical hypotheses can animate interest in STEM (Science, Innovation, Designing, and Math) professions and exploration.

- Social and Philosophical Implications: The hypothesis could likewise have social and philosophical ramifications, testing how we might interpret our spot in the universe and the idea of astronomical cycles. It welcomes reflection on the more extensive inquiries regarding the starting points of heavenly bodies and the instruments overseeing their arrangement.

- Effect on Space Exploration: Assuming that Howard's hypothesis builds up forward momentum, it could impact future space investigation missions. Space offices could plan missions to test parts of the hypothesis, for example, dissecting sun based material or researching planetary development processes. This could extend our insight into the nearby planet group and then some.

CHAPTER 7:

Comparative Theories

To assess Terrence Howard's hypothesis that the Sun is the beginning, everything being equal, contrasting it and different speculations of planetary formation is fundamental. This part analyzes a few laid out and arising hypotheses, featuring their similitudes, contrasts, and how they address the development of planetary frameworks.

7.1 The Protoplanetary Plate Model

- Overview: The protoplanetary plate model is the most broadly acknowledged hypothesis of planetary development. It recommends that planets structure from an alternating circle of gas and residue encompassing a youthful star. After some time, particles in the plate impact and remain together, shaping bigger bodies through a cycle called growth. These bodies in the end mix into planets.

- Mechanisms: As per this model, the circle's material is made out of remainders from the atomic cloud that shaped the star. The arrangement cycle includes the progressive aggregation of material in the circle, prompting the development of planetesimals and protoplanets.

- Evidence: Observational proof supporting the protoplanetary plate model incorporates pictures of circles around youthful stars, for example, those caught by the Hubble Space Telescope and the Atacama Enormous Millimeter/submillimeter Cluster (ALMA). These perceptions show the construction and sythesis of circles and their job in planetary arrangement.
- Examination with Howard's Theory: Howard's hypothesis challenges the protoplanetary circle model by proposing that the actual Sun creates planets, as opposed to the planets shaping from an encompassing plate. The protoplanetary circle model is upheld by broad observational information, while Howard's hypothesis requires a re-assessment of the cycles by which planets are shaped.

7.2 The Nebular Hypothesis

- Overview: The nebular speculation is a previous hypothesis that originates before the protoplanetary plate model. It recommends that the planetary group framed from a goliath haze of gas and residue. As the cloud fell under gravity, it turned quicker and smoothed into a circle, with the Sun framing at the middle and the planets shaping from the leftover material in the plate.

- Mechanisms: This speculation underscores the job of gravitational breakdown in shaping a protoplanetary circle. The plate's material slowly amassed together to shape planetesimals, which then, at that point, mixed into planets.

- Evidence: The nebular speculation is upheld by perceptions of star-shaping locales and the disclosure of protoplanetary plates. Present day varieties of the nebular speculation adjust intimately with the protoplanetary circle model.

- Correlation with Howard's Theory: Both the nebular speculation and Howard's hypothesis include the Sun as a focal component in planetary development, yet they vary in their systems. While the nebular speculation upholds the possibility of an encompassing circle, Howard's hypothesis proposes direct material emanation from the Sun.

7.3 The Center Gradual addition Model

- Overview: The center growth model is a refinement of the protoplanetary circle model. It proposes that planets structure by accumulating strong material to develop a center, which then, at that point, draws in a gas envelope to frame gas monsters or stays an earthly planet.

- Mechanisms: As per this model, little particles in the protoplanetary plate impact and stay together to frame planetesimals. These planetesimals then, at that point, crash to frame protoplanets. Once a protoplanet arrives at a minimum amount, it can draw in a lot of gas from the circle, shaping a gas monster.

- Evidence: The center gradual addition model is upheld by perceptions of exoplanetary frameworks and recreations of planetary development. The presence of both earthbound and gas goliath exoplanets gives proof to this model.

- Examination with Howard's Theory: The center growth model spotlights on the slow development of planetary centers from strong material in a circle, while Howard's hypothesis suggests that the actual Sun is the wellspring of planetary material. Howard's hypothesis would have to represent the definite cycles of center arrangement and gas gradual addition portrayed by the center accumulation model.

7.4 The Plate Unsteadiness Model

- Overview: The plate unsteadiness model is an elective hypothesis for shaping gas goliath planets. It suggests that planets structure through gravitational hazards inside the protoplanetary circle, prompting the quick breakdown of districts inside the plate to shape gas monsters.

- Mechanisms: As per this model, certain areas of the protoplanetary circle become gravitationally unsteady and breakdown to straightforwardly frame gas monster planets. This cycle is quicker than the center growth model and doesn't depend on the steady development of a strong center.

- Evidence: Observational proof supporting the plate shakiness model incorporates the disclosure of youthful, gigantic gas goliaths in star-framing areas. These perceptions recommend that gas goliaths can frame rapidly through plate hazards.

- Correlation with Howard's Theory: The plate insecurity model spotlights on processes inside the protoplanetary circle to frame gas goliaths, though Howard's hypothesis recommends that the actual Sun produces planets. The circle flimsiness model's accentuation on quick development stands out from the progressive cycles proposed by Howard's hypothesis.

7.5 The Sun oriented Cloud Theory

- Overview: The sun oriented cloud hypothesis is a variety of the nebular speculation, stressing the job of rakish force and the preservation of mass in the development of the planetary group. It portrays how the sun powered cloud developed into a protoplanetary plate, with planets framing from the circle material.

- Mechanisms: This hypothesis consolidates the protection of precise force and depicts how the sun oriented cloud's turn prompted the development of a smoothed plate. Planets framed through growth of material inside this circle.

- Evidence: The sun oriented cloud hypothesis is upheld by perceptions of protoplanetary plates and the construction of the planetary group. It gives a definite clarification of the elements engaged with plate development and planetary growth.

- Correlation with Howard's Theory: The sun oriented cloud hypothesis lines up with the protoplanetary circle model and gives a deeply grounded structure to figuring out planetary development. Howard's hypothesis wanders by suggesting that planets start straightforwardly from the Sun as opposed to from an encompassing plate.
- Half breed Models: A few arising speculations endeavor to join components from various models, for example, consolidating parts of both the protoplanetary plate model and the circle unsteadiness model. These mixture approaches try to address the limits of individual speculations and give a more far reaching comprehension of planetary development.

- Elective Mechanisms: New examination keeps on investigating elective components for planetary development, including the job of attractive fields, choppiness, and substance processes in the arrangement and advancement of planetary frameworks.

- Observational Advances: Advances in observational innovation and information examination might prompt new experiences into planetary arrangement processes. Future disclosures could challenge existing models and add to the advancement of new hypotheses.

CHAPTER 8:

Theoretical Models and Simulations

The investigation of Terrence Howard's hypothesis — that the actual Sun is the beginning of all planets — requires hearty hypothetical models and reenactments to evaluate its practicality and suggestions. This section dives into the hypothetical systems and computational reproductions used to assess planetary arrangement speculations, including Howard's speculation, and examines how these instruments can be utilized to investigate and test offbeat thoughts.

8.1 Hypothetical Models in Planetary Formation

Hypothetical models are fundamental for understanding the cycles associated with planetary development. They give a structure to mimicking the circumstances and components that lead to the development of planetary frameworks. These models differ in intricacy and degree, it being explored to rely upon the particular cycles.

8.1.1 The Protoplanetary Plate Model

- Framework: The protoplanetary circle model is grounded in the idea that planets structure from an alternating plate of gas and residue around a youthful star. The model integrates the physical science of circle elements, gradual addition, and rakish force preservation. It portrays how material in the plate bit by bit bunches together to frame planetesimals and protoplanets.

- Key Equations: The model includes conditions administering circle advancement, for example, the Navier-Stirs up conditions for liquid elements, and conditions portraying the development of strong particles. The equilibrium of powers, including gravity, pressure, and gooey powers, assumes a significant part in plate conduct and planetary development.

- Applications: Recreations in view of this model can foresee the appropriation of material in the plate, the development paces of planetesimals, and the arrangement of planetary frameworks. They can likewise investigate different starting circumstances and plate properties to comprehend what they mean for planetary arrangement results.

8.1.2 The Center Gradual addition Model

- Framework: The center gradual addition model spotlights on the arrangement of planetary centers through the growth of strong material in the protoplanetary plate. When a center arrives at a minimum amount, it can draw in a vaporous envelope to frame a gas goliath. This model coordinates cycles like center development, gas growth, and relocation.

- Key Equations: The model includes conditions for center development rates, gas accumulation rates, and the elements of planetesimal impacts. It additionally considers the impact of gas drag and the connection between the center and the encompassing circle.

- Applications: Reenactments utilizing the center accumulation model can foresee the qualities of various kinds of planets, like earthly planets and gas monsters. They can likewise research how center arrangement and gas gradual addition rely upon plate properties and introductory circumstances.

8.1.3 The Circle Insecurity Model

- Framework: The plate unsteadiness model recommends that gas monster planets structure through gravitational dangers inside the protoplanetary circle. Locales of the plate become gravitationally unsound and breakdown to quickly frame gas monsters. This model underscores the job of circle disturbance and gravitational powers.

- Key Equations: The model consolidates conditions for circle strength, gravitational breakdown, and the development of thick clusters. It likewise includes estimations of gas elements and the circumstances expected for insecurity to happen.

- Applications: Reproductions in view of the circle shakiness model can investigate the circumstances under which gas goliath development is leaned toward. They can likewise break down how plate properties, like temperature and thickness, impact the probability of flimsiness and fast planet development.

8.2 Reenactments and Computational Approaches

Reenactments are significant for testing hypothetical models and investigating the unique cycles associated with planetary arrangement. They permit scientists to display complex cooperations and survey the plausibility of various speculations, including Howard's theory.

8.2.1 Mathematical Simulations

- Methods: Mathematical reproductions include settling the overseeing conditions of planetary development utilizing computational techniques. Procedures like limited distinction techniques, limited component strategies, and smoothed molecule hydrodynamics (SPH) are regularly used to reenact liquid elements, plate development, and gradual addition processes.

- Challenges: Mimicking planetary development includes managing many spatial and fleeting scales. Scientists should adjust exactness and computational effectiveness, frequently utilizing superior execution figuring assets to deal with the intricacy of recreations.

- Applications: Mathematical reenactments can show the development and advancement of planetary frameworks, including the development of planetesimals, the elements of protoplanetary plates, and the collaborations between shaping planets and the encompassing circle. They can likewise test the expectations of various hypotheses against observational information.

8.2.2 Hydrodynamic Simulations

- Methods: Hydrodynamic reproductions center around the way of behaving of liquids, remembering gas and residue for protoplanetary circles. These reenactments address the conditions of liquid elements to demonstrate the stream, disturbance, and collaborations inside the circle.

- Challenges: Hydrodynamic reenactments should represent different actual cycles, like choppiness, consistency, and attractive fields. They require precise displaying of the circle's properties and the collaborations between various parts.

- Applications: Hydrodynamic reproductions can investigate what disturbance and plate structure mean for planet arrangement. They can likewise explore the job of attractive fields and different elements in molding the circle and affecting planetary growth.

8.2.3 N-Body Simulations

- Methods: N-body reenactments center around the collaborations between different bodies, like planetesimals and protoplanets. These reproductions utilize gravitational elements to demonstrate the development of planetary frameworks and the arrangement of planets through impacts and accumulation.

- Challenges: N-body reproductions require precise demonstrating of gravitational powers and the elements of various bodies. They can be computationally escalated, particularly while demonstrating huge quantities of bodies overstretched periods.

- Applications: N-body reenactments can explore the development of planetary frameworks, including the development and relocation of planets. They can likewise investigate the elements of crashes and the impacts of resonances and connections among planets and different bodies.

8.3 Assessing Howard's Hypothesis with Simulations

To evaluate Terrence Howard's hypothesis, recreations should investigate the proposed instruments of sun oriented material discharge and planetary development. This includes a few key stages:

8.3.1 Growing New Models

- Framework: Analysts need to foster new hypothetical models that consolidate Howard's theory. This includes characterizing how the Sun could discharge material that structures planets and how this material would develop into planetary bodies.

- Key Equations: New conditions should be formed to depict the cycles of sun oriented material outflow, material elements, and planetary growth. These conditions ought to coordinate with existing models of heavenly and planetary material science.

8.3.2 Reenacting Sun oriented Outflow and Accretion

- Methods: Reenactments should show the outflow of material from the Sun and its resulting elements. This incorporates recreating the vehicle, conveyance, and gradual addition of catapulted material to shape planets.

- Challenges: Mimicking sun based discharge includes complex cooperations between the Sun's action and the encompassing space climate. Scientists should consider the impacts of sun based flares, coronal mass discharges, and the scattering of material.

8.3.3 Contrasting and Observational Data

- Methods: Reenactments ought to be contrasted with observational information with test Howard's hypothesis. This incorporates examining perceptions of sun powered action, planetary development processes, and the creation of planetary frameworks.

- Challenges: Contrasting reproductions and perceptions requires precise information and cautious translation. Scientists should evaluate whether the recreated results line up with certifiable perceptions and how they match the expectations of Howard's hypothesis.

8.4 Future Headings in Demonstrating and Simulation

The investigation of planetary arrangement is consistently advancing, and new advancements in displaying and reproduction methods offer energizing prospects:

- Progressions in Computational Power: Proceeded with upgrades in computational assets will empower more itemized and exact reenactments of planetary development processes. Improved computational power will work with the investigation of additional complicated models and bigger scope reproductions.

- Joining of Machine Learning: AI methods can be applied to examine reenactment information, recognize designs, and enhance model boundaries. Incorporating AI with customary recreations can improve our capacity to investigate and approve new hypotheses.

- Cooperative Research: Cooperative endeavors between specialists, organizations, and space organizations can propel the investigation of planetary arrangement. Joint activities can consolidate skill and assets to foster new models, direct recreations, and test capricious hypotheses.

- Imaginative Observations: Future perceptions with cutting edge telescopes and space missions will give new information to refine models and test speculations. Developments in observational innovation will upgrade our capacity to investigate the cycles of planetary arrangement and test speculations.

CHAPTER 9:

Scientific and Philosophical Perspectives

Terrence Howard's hypothesis that the actual Sun is the beginning of all planets challenges existing logical standards as well as brings up significant philosophical issues about how we might interpret the universe. This part investigates both the logical and philosophical ramifications of Howard's speculation, looking at how it fits inside the more extensive setting of logical request and human idea.

9.1 Logical Perspectives

9.1.1 Reconsidering Laid out Models

Howard's hypothesis requires a re-assessment of laid out models of planetary development. The dominating models, for example, the protoplanetary plate model and the center gradual addition model, depend on broad observational information and hypothetical turn of events. Howard's speculation presents another system for planetary development, recommending that planets structure straightforwardly from material launched out by the Sun.

- Influence on Astrophysics: Assuming that approved, Howard's hypothesis could prompt huge amendments in astrophysical models. Specialists would have to consolidate new components for sun based material outflow and planetary development, possibly prompting a more far reaching comprehension of heavenly and planetary frameworks.

- Coordination with Existing Data: The test lies in accommodating Howard's hypothesis with existing observational information. While conventional models are upheld by broad proof, Howard's speculation would require new information and recreations to confirm its legitimacy and incorporate it with current information.

9.1.2 Testing and Validation

Testing Howard's hypothesis includes thorough logical techniques, including:

- Observational Evidence: Assembling new observational information to test the expectations of Howard's hypothesis. This incorporates examining sunlight based material and its likely job in planetary development, as well as noticing other star frameworks to check whether comparable cycles happen.

- Recreations and Models: Creating and running reproductions to investigate the systems proposed by Howard's hypothesis. Contrasting these reproductions and noticed peculiarities will assist with evaluating the hypothesis' possibility and exactness.

- Peer Audit and Debate: Drawing in with mainstream researchers through peer survey and scholastic discussion. This interaction is fundamental for basically assessing Howard's hypothesis, tending to likely shortcomings, and refining the speculation in light of criticism and new discoveries.

9.1.3 Ramifications for Related Fields

- Sun oriented Physics: Howard's hypothesis could have suggestions for sun based physical science, especially in figuring out sun powered action and its consequences for the general climate. This incorporates investigating how sun based material discharge could impact planetary arrangement and planetary group elements.

- Planetary Science: The hypothesis may likewise affect planetary science, particularly in figuring out the organization and arrangement of planets. Analysts would have to explore how sunlight based material could add to the development of various sorts of planets and their qualities.

9.2 Philosophical Perspectives

9.2.1 Idea of Origins

Howard's hypothesis challenges conventional ideas of infinite beginnings and the development of planetary frameworks. Logically, it brings up issues about the idea of creation and the cycles that administer the universe.

- Enormous Genesis: The possibility that the actual Sun could be the beginning of planets challenges the idea of an inactive star and presents a more unique perspective on grandiose beginning. It prompts a reconsideration of how heavenly bodies appear and the job of stars in molding planetary frameworks.

- Human-centered Considerations: Philosophical reflections on the ramifications for human comprehension of our spot in the universe. On the off chance that Howard's hypothesis were approved, it would recommend a more interconnected connection among stars and planets, impacting our impression of inestimable cycles and human importance.

9.2.2 Logical Standards and Innovation

- Worldview Shifts: Howard's speculation addresses a potential change in perspective in stargazing and astronomy. It features the job of unpredictable thoughts in progressing logical information and testing laid out ideal models.

- Logical Creativity: The hypothesis highlights the significance of inventiveness and creative mind in logical request. It exhibits how investigating clever thoughts can prompt new experiences and drive logical advancement, regardless of whether they at first appear to be capricious.

9.2.3 Ramifications for Information and Understanding

- Epistemological Questions: The hypothesis brings up epistemological issues about how information is procured and approved. It moves specialists to consider how new speculations are assessed and acknowledged inside mainstream researchers.

- Reasoning of Science: The hypothesis adds to the way of thinking of science by showing the cycles of hypothesis advancement, testing, and acknowledgment. It features the job of proof, distrust, and decisive reasoning in forming logical comprehension.

9.2.4 Moral and Cultural Considerations

- Moral Implications: While Howard's hypothesis doesn't straightforwardly include moral issues, it prompts reflection on the moral obligations of researchers in investigating and conveying novel thoughts. This incorporates guaranteeing that hypotheses are thoroughly tried and that discoveries are precisely announced.

- Cultural Impact: The hypothesis' effect on open comprehension of science and space investigation. If broadly acknowledged, it could impact instructive educational programs, public view of cosmology, and interest in space research. It underlines the requirement for powerful science correspondence and commitment with people in general.

9.3 The Job of Interdisciplinary Approaches

- Coordinating Perspectives: Understanding Howard's hypothesis benefits from interdisciplinary methodologies that join bits of knowledge from astronomy, theory, and different fields. Joint efforts between researchers, scholars, and teachers can upgrade how we might interpret the hypothesis' suggestions and encourage a more all encompassing perspective on planetary development.

- Expanding Inquiry: Interdisciplinary exploration can widen the extent of request, prompting new bits of knowledge and advancements. By coordinating alternate points of view, scientists can foster more complete models and investigate unusual thoughts from numerous points.

9.4 Future Philosophical and Logical Inquiry

- Continuous Exploration: The investigation of Howard's hypothesis is a continuous interaction that includes both logical examination and philosophical reflection. Future examination will keep on tending to the hypothesis' legitimacy, suggestions, and effect on how we might interpret the universe.

- Empowering Open Inquiry: The hypothesis features the significance of encouraging an open and curious way to deal with logical exploration. Empowering interest and investigation of eccentric thoughts can prompt new revelations and progressions in information.

CHAPTER 10:

The Role of Media and Public Perception

The scattering of logical speculations and theories, including Terrence Howard's thought that the Sun is the beginning, everything being equal, is essentially affected by media and public insight. This section inspects what media inclusion and public figuring out mean for the acknowledgment and effect of such speculations, investigating the elements of correspondence between researchers, the media, and the general population.

10.1 Media Portrayal of Logical Theories

10.1.1 Media Inclusion and Sensationalism

- Sensationalism: The media frequently sensationalizes logical speculations to catch public consideration. This can prompt overstated or contorted portrayals of complicated thoughts. Howard's hypothesis, being unusual, is helpless to dramatist depictions that may not precisely mirror the logical subtleties of the speculation.

- Impact: Dramatist inclusion can produce interest and energy however may likewise prompt false impressions or distortions. It is fundamental for researchers to resolve these issues by giving clear, exact clarifications and drawing in with the media to address any mistakes.

10.1.2 Media Channels and Formats

- Conventional Media: Papers, magazines, and TV have generally been the essential wellsprings of logical data. Inclusion in these organizations can shape public discernment through articles, narratives, and news portions.

- Computerized Media: The ascent of advanced media, including sites, virtual entertainment, and online news stages, has sped up and expansiveness of data dispersal. While advanced media considers more extensive reach and more intuitive commitment, it additionally presents difficulties connected with the exactness and believability of data.

- Job of Science Communicators: Science communicators, including writers and advertising specialists, assume an essential part in making an interpretation of mind boggling logical speculations into open language. Powerful correspondence is fundamental for keeping up with public interest and understanding while at the same time guaranteeing exact portrayal.

10.2 Public Impression of Eccentric Theories

10.2.1 Acknowledgment and Skepticism

- Introductory Reaction: Whimsical hypotheses, like Howard's, frequently face suspicion from mainstream researchers and general society. Introductory responses can be impacted by earlier information, individual convictions, and the apparent validity of the source.

- Impact of Authority: Public discernment is vigorously affected by the assessments of laid out specialists and organizations. Support or investigate from trustworthy researchers can fundamentally influence how a hypothesis is gotten by general society.

10.2.2 Schooling and Understanding

- Logical Literacy: Public comprehension of logical hypotheses is frequently connected to logical proficiency and schooling. More significant levels of instruction by and large connect with better cognizance of mind boggling thoughts and hypotheses.

- Instructive Outreach: Effort programs, instructive materials, and public talks can assist with overcoming any barrier between logical examination and public comprehension. Drawing in with schools, colleges, and local area associations can advance informed conversations and upgrade logical education.

10.3 The Effect of Media on Logical Discourse

10.3.1 Forming Public Debate

- Public Discourse: Media inclusion impacts public discussion about logical hypotheses by outlining issues, featuring contentions, and molding stories. Howard's hypothesis, if generally covered, could turn into a subject of conversation in more extensive logical and public gatherings.

- Effect on Policy: Public discernment, molded by media inclusion, can likewise impact science strategy and financing choices. High-profile speculations and discussions can stand out from policymakers and subsidizing organizations, influencing the course of exploration and asset portion.

10.3.2 The Criticism Loop
- Media Effect on Research: Media inclusion can affect logical examination by causing to notice explicit hypotheses and impacting research needs. The perceivability of Howard's hypothesis could prompt expanded research center around sun oriented and planetary arrangement.

- Research Impact on Media: On the other hand, headways and discoveries from logical exploration can drive media accounts. Positive or negative advancements in research connected with Howard's hypothesis might impact how the media provides details regarding and talks about the speculation.

10.4 Systems for Powerful Science Communication

10.4.1 Clear and Precise Reporting

- Keeping away from Misrepresentation: Researchers and communicators should cooperate to guarantee that speculations are accounted for precisely. Clear, exact language and setting are fundamental for forestalling error and falsehood.

- Drawing in with Media: Laying out proactive associations with columnists and news sources can assist with guaranteeing exact and adjusted inclusion. Researchers ought to be ready to make sense of complicated ideas in available terms and address any misinterpretations.

10.4.2 Public Commitment and Dialogue

- Intuitive Platforms: Using intelligent stages, like virtual entertainment and public discussions, considers continuous commitment with general society. These stages can work with conversations, answer questions, and give reports on continuous examination.

- Instructive Initiatives: Advancing science training and proficiency through open talks, instructive projects, and media content can improve public comprehension and encourage informed conversations about logical speculations.

10.5 Contextual analyses and Examples
10.5.1 Prominent Logical Theories

- Contextual analysis: The Heliocentric Model: The authentic acknowledgment of the heliocentric model by Copernicus and Galileo shows the job of media and public discernment in the acknowledgment of new hypotheses. The difficulties looked by these early hypotheses feature the significance of proof and correspondence in beating suspicion.

- Contextual investigation: The Higgs Boson Discovery: The disclosure of the Higgs boson and its media inclusion show how complex logical revelations are imparted to general society. The cycle included broad media commitment, state funded instruction, and the utilization of clear, available language.

10.5.2 Arising Theories

- Contextual analysis: Exoplanet Discoveries: The disclosure of exoplanets has been broadly canvassed in the media, affecting public discernment and interest in space investigation. The depiction of these revelations has added to a more extensive comprehension of planetary frameworks and the quest for livable universes.

10.6 Future Directions

- Creative Specialized Methods: Future improvements in media and correspondence innovation will keep on molding how logical speculations are introduced and seen. Embracing new instruments and techniques can improve the adequacy of science correspondence and public commitment.

- Cooperative Efforts: Cooperation between researchers, communicators, and media experts is critical for successful science correspondence. Building associations and encouraging open exchange can assist with tending to difficulties and advance exact, informed conversations about logical speculations.

CONCLUSION

Terrence Howard's hypothesis that the Sun is the beginning of all planets addresses a huge takeoff from laid out galactic ideal models. By recommending that the actual Sun could emanate material that in the long run combines into planets, Howard challenges ordinary models of planetary development, for example, the protoplanetary plate and center accumulation models. This hypothesis welcomes a reconsideration of how we might interpret heavenly and planetary elements, requiring thorough logical assessment and philosophical reflection.

Logical Evaluation

The legitimacy of Howard's hypothesis relies on its capacity to endure examination through hypothetical models and observational proof. Existing models of planetary development give a system to understanding how planets rise up out of protoplanetary plates or through gravitational precariousness. Howard's speculation presents an original component that requires the improvement of new hypothetical models and recreations to investigate its practicality. The progress of these models in anticipating perceptible peculiarities and lining up with exact information will be significant for surveying the hypothesis' legitimacy.

Reenactments and computational methodologies are indispensable for investigating the mechanics of Howard's speculation. Mathematical reenactments, hydrodynamic models, and N-body recreations offer devices to explore the likely cycles by which sun oriented material could add to planetary development. These reenactments should be contrasted with observational information with test the forecasts of the hypothesis and refine how we might interpret sun powered and planetary associations.

Philosophical Implications

Howard's hypothesis stretches out past the domain of science into philosophical contemplations about the idea of enormous starting points and our spot in the universe. The speculation challenges customary perspectives on heavenly and planetary development, provoking a reexamination of how divine bodies appear. Rationally, it brings up issues about the idea of logical advancement, the job of imagination in logical request, and the manners by which new hypotheses can move how we might interpret the universe.

The hypothesis additionally features the powerful interchange among logical and philosophical request. The potential change in outlook presented by Howard's speculation highlights the significance of thinking about both logical proof and philosophical appearance in propelling our insight. It shows the way that eccentric thoughts can prompt new experiences and cultivate a more nuanced comprehension of the universe.

Media and Public Perception

The job of media and public discernment is urgent in forming the talk around logical hypotheses. Dramatist inclusion and the impact of media on open comprehension can affect the acknowledgment and scattering of groundbreaking thoughts. Successful science correspondence is fundamental for guaranteeing exact portrayal of speculations and encouraging informed public conversations.

Howard's hypothesis, if generally covered, could create huge public interest and discussion. The media's depiction of the speculation will impact the way things are seen by both established researchers and the overall population. Guaranteeing clear, precise detailing and drawing in with people in general through instructive drives can assist with spanning holes in understanding and address expected confusions.

Future Directions

As examination proceeds, the investigation of Howard's hypothesis will profit from continuous logical request, interdisciplinary coordinated effort, and headways in innovation. Future examination ought to zero in on creating and refining hypothetical models, directing reproductions, and dissecting observational information to assess the plausibility of the speculation.

The coordination of new specialized techniques and the advancement of science training will likewise assume an essential part in forming public discernment and encouraging informed conversations about the hypothesis. Embracing creative ways to deal with science correspondence and empowering open exchange between researchers, communicators, and the public will be fundamental for propelling comprehension we might interpret planetary development and the more extensive ramifications of Howard's hypothesis.

In synopsis, Terrence Howard's hypothesis addresses a strong and unpredictable way to deal with grasping the starting points of planets. Its assessment requires a mix of thorough logical examination, philosophical reflection, and viable correspondence. By investigating the logical, philosophical, and public elements of the hypothesis, we gain a more profound enthusiasm for its possible effect and the more extensive journey for information about our universe.

APPENDICES

Supplement A: Key Phrasing and Definitions

1. Protoplanetary Disk: An alternating plate of thick gas and residue encompassing a youthful star, from which planets and other divine bodies structure.

2. Core Growth Model: A hypothesis proposing that planets structure by the progressive gathering of strong material to construct a

center, which then, at that point, draws in an encompassing gas envelope.

3. Disk Unsteadiness Model: A hypothesis recommending that gas goliath planets structure through the gravitational breakdown of thick locales inside the protoplanetary circle.

4. Hydrodynamics: The investigation of liquids moving, including gases and fluids, which is pivotal for understanding the elements of protoplanetary circles and planetary arrangement.

5. N-Body Simulation: A computational strategy used to display the gravitational collaborations between different heavenly bodies, like planetesimals and protoplanets.

6. Sensationalism: The act of introducing data in a way that is intended to incite public interest or fervor, frequently to the detriment of exactness.

7. Scientific Literacy: The capacity to comprehend and apply logical ideas and cycles, which impacts how people decipher and draw in with logical speculations.

Addendum B: Synopsis of Key Speculations in Planetary Formation

1. Protoplanetary Circle Theory: Proposes that planets structure from an alternating plate of gas and residue around a youthful star. Planetesimals inside the circle impact and blend to frame protoplanets, which then develop into planets.

2. Core Growth Theory: Recommends that planets structure by the continuous collection of strong material to make a center. When the center arrives at a minimum amount, it draws in a gas envelope to frame a gas monster.

3. Disk Shakiness Theory: Proposes that gas monster planets structure through the gravitational flimsiness of locales inside the

protoplanetary plate. These districts breakdown quickly to frame planets.

4. Solar Cloud Theory: An expansion of the protoplanetary plate hypothesis that consolidates the sunlight based cloud's part in molding the arrangement of the planetary group's planets and different bodies.

Index C: Strategies for Reenacting Planetary Formation

1. Numerical Simulations: Utilize computational calculations to address the conditions administering circle elements, gradual addition processes, and planetary arrangement. These reproductions can demonstrate a large number of situations and starting circumstances.

2. Hydrodynamic Simulations: Spotlight on the way of behaving of gases and liquids inside the protoplanetary circle. They model how disturbance, consistency, and different elements

impact the plate's advancement and planet arrangement.

3. N-Body Simulations: Model the gravitational communications between different heavenly bodies. These recreations are utilized to concentrate on the elements of planetesimal impacts, the development of protoplanets, and the development of planetary frameworks.

4. Monte Carlo Methods: Factual techniques used to reenact the impacts of irregularity and vulnerability in planetary development processes. They give bits of knowledge into the probability of different results in light of probabilistic models.

Reference section D: Remarkable Figures and Diagrams

1. Protoplanetary Circle Structure: A chart outlining the run of the mill construction of a protoplanetary plate, including the focal star,

plate districts, and the development zones for planetesimals and protoplanets.

2. Core Growth Process: A schematic appearance the means engaged with the center growth process, from the collection of strong material to the development of a gas goliath planet.

3. Disk Unsteadiness Model: A portrayal of the plate insecurity model, featuring how gravitational breakdown inside the circle prompts the development of gas monster planets.

4. Simulation Results: Diagrams and pictures from mathematical and hydrodynamic recreations showing the advancement of protoplanetary plates and the arrangement of planetary bodies.

Reference section D: Remarkable Figures and Diagrams

1. Protoplanetary Circle Structure: A chart outlining the run of the mill construction of a protoplanetary plate, including the focal star, plate districts, and the development zones for planetesimals and protoplanets.

2. Core Growth Process: A schematic appearance the means engaged with the center growth process, from the collection of strong material to the development of a gas goliath planet.

3. Disk Unsteadiness Model: A portrayal of the plate insecurity model, featuring how gravitational breakdown inside the circle prompts the development of gas monster planets.

4. Simulation Results: Diagrams and pictures from mathematical and hydrodynamic recreations showing the advancement of protoplanetary plates and the arrangement of planetary bodies.

Reference section E: Extra Resources

1. Research Papers and Articles: A rundown of key examination papers and articles pertinent to planetary development hypotheses, including both primary investigations and ongoing headways.

2. Books and Textbooks: Suggested readings for a more profound comprehension of planetary development, heavenly physical science, and computational demonstrating.

3. Online Data sets and Tools: Connections to online information bases, recreation devices, and programming utilized for demonstrating planetary arrangement and examining observational information.

4. Educational Websites: Assets for teachers and understudies keen on looking further into planetary science, astronomy, and logical reenactments.

Addendum F: Glossary of Terms

- Accretion: The cycle by which material gathers through gravitational fascination or crashes.

- Precise Momentum: The amount of pivot of a body, which influences the elements of protoplanetary plates and planetary development.

- Planetesimal: A little body framed from the residue and gas in a protoplanetary plate, which can impact and mix to shape bigger planetary bodies.

- Sun powered Nebula: The haze of gas and residue from which the nearby planet group shaped, enveloping the early protoplanetary plate.

- Heavenly Dynamics: The investigation of the movement and collaboration of stars and other divine bodies, remembering their job for planetary development.

ACKNOWLEDGEMENTS

The improvement of this book on Terrence Howard's hypothesis of the Sun as the beginning, all things considered, could never have been conceivable without the commitments and backing of numerous people and foundations.

Above all else, I stretch out my most profound appreciation to Terrence Howard for rousing this investigation into an unusual and provocative speculation. His strong thoughts and ability to challenge laid out ideal models have ignited huge conversations and request in the field of astronomy.

I'm significantly appreciative to the intellectual and exploration networks that have given priceless experiences and input. Exceptional thanks to the specialists in astronomy, planetary science, and computational demonstrating who have shared their insight and points of view, which have been fundamental in contextualizing and assessing Howard's hypothesis. Their thorough investigations and useful studies have advanced this work.

A genuine thank you goes to the essayists, editors, and distributers who have carried this book to completion. Their impressive skill, devotion, and mastery in refining the substance and guaranteeing its lucidity and openness have been significant.

To the establishments and exploration associations that gave assets and backing to this undertaking, including admittance to scholastic papers, recreations, and data sets, I offer my true appreciation. Their commitments have extraordinarily worked with the top to bottom examination and investigation introduced in this book.

I'm likewise thankful to the media and science communicators who take care of and talked about Howard's hypothesis. Their job in scattering logical thoughts and drawing in general society has had a significant impact in molding the talk around this speculation.

In conclusion, I might want to recognize the help of loved ones, whose consolation and understanding have been a wellspring of inspiration all through the creative cycle. Their understanding and confidence in this undertaking have been significant.

Much obliged to you to all who have added to the advancement of this book and to the continuous journey for figuring out in the field of astronomy. Your commitments have made this investigation of Terrence Howard's hypothesis conceivable and significant.

www.ingramcontent.com/pod-product-compliance
Lightning Source LLC
Chambersburg PA
CBHW072052230526
45479CB00010B/681